YOUR KNOWLEDGE HAS VALUE

- We will publish your bachelor's and master's thesis, essays and papers

- Your own eBook and book - sold worldwide in all relevant shops

- Earn money with each sale

Upload your text at www.GRIN.com
and publish for free

Bibliographic information published by the German National Library:

The German National Library lists this publication in the National Bibliography; detailed bibliographic data are available on the Internet at http://dnb.dnb.de .

This book is copyright material and must not be copied, reproduced, transferred, distributed, leased, licensed or publicly performed or used in any way except as specifically permitted in writing by the publishers, as allowed under the terms and conditions under which it was purchased or as strictly permitted by applicable copyright law. Any unauthorized distribution or use of this text may be a direct infringement of the author s and publisher s rights and those responsible may be liable in law accordingly.

Imprint:

Copyright © 2015 GRIN Verlag, Open Publishing GmbH
Print and binding: Books on Demand GmbH, Norderstedt Germany
ISBN: 978-3-668-05315-1

This book at GRIN:

http://www.grin.com/en/e-book/306482/environmental-policy-in-the-us-the-clean-water-act-and-global-warming

Florence Mwangi

Environmental Policy in the US. The Clean Water Act and Global Warming

GRIN Publishing

GRIN - Your knowledge has value

Since its foundation in 1998, GRIN has specialized in publishing academic texts by students, college teachers and other academics as e-book and printed book. The website www.grin.com is an ideal platform for presenting term papers, final papers, scientific essays, dissertations and specialist books.

Visit us on the internet:

http://www.grin.com/

http://www.facebook.com/grincom

http://www.twitter.com/grin_com

Environmental Policy in the US
The Clean Water Act and Global Warming

Florence Mwangi

Table of Contents

1. U.S. environmental law (Clean Water Act) 2
2. The economic impact of the Clean Water Law. 4
3. Improvement of Clean Water Act on the environment or situation 4
4. Do you think that sound science has proven that global warming is a credible threat or not? 5
5. Should the United States adopt additional policies or laws to curb greenhouse gas emissions? 6
6. Conclusion 7
7. Works Cited 8

This paper was written by a non-native speaker of English.
Please excuse any linguistic mistakes or inconsistencies.

1. **U.S. environmental law (Clean Water Act)**

It develops the basic structure for regulating discharge of major pollutants in the water and controlling quality standards of the surface water. The Act was first enacted in 1948 and was then referred to as Water Pollution Control Act. It was with time significantly reorganized and amended in 1972. These amendments gave the Act its present form and developed a national goal and objectives that all waters of the United State shall be fishable as well as swimmable. The set goals were to be achieved through elimination of all pollutant discharge in to the water by 1985 with making water for aquatic life and human beings by 1st July 1983 being an interim goal (History of the clean water, 2013).

Further amendments were done in 1977 changing the initial title of the Act to its current title i.e. Clean Water Act of 1977. Additional amendments were enacted in 1981 (Municipal Wastewater Treatment Construction Grant Amendment) and in 1987 was the Water Quality Act. The act regulates discharge to water through permits issued under the National Pollutant Discharge Elimination System permitting Program. These permits are issued by The division of Water Quality Protection while the Water Enforcement Branch assures ensures that all the discharges comply with the NPDES permits (History of the clean water, 2013).

Section 301 of the Clean Water Act prohibits discharge of pollutants into water from the point sources without obtaining the permit. Some of the requirements of the permit were treatment of the pollutant to the extent that will comply with established water quality standards. Examples of the discharges that were unlawful without the permit are:

i. Car wash, power washer and carpet cleaners where wash water flow into the storm drain.
ii. Storm water runoff from the industrial activities.
iii. Animal feedlots lagoon discharge into rivers and streams.

iv. Dumping of hazardous materials into a wetland.

v. Piping the sanitary drain direct into lake.

vi. Industrial process water into the water sources.

vii. Discharging of municipality wastewater from treatment facility.

The table below shows a summary of the major provisions of the Clean Water Act and amendments done over the time.

TITLE	YEAR OF ENACTMENT	KEY PROVISIONS
Water pollution act	1948	Federal research authorization
Water pollution Act Amendment	1956	State responsibility to develop water quality criteria
		Discretionary federal responsibility to convene enforcement conferences
Water Quality Act	1965	Ambient standards
		State enforcement and implementation responsibilities
Federal Water Pollution Control Act	1972	Conventional pollutants technology based standards
		Federal implementation and enforcement responsibility
Coastal Zone Management Act	1972	Directed towards coastal water pollution issues
Clean Water Act Amendment	1977	Increased coverage to toxic pollutants
Municipal Wastewater Treatment Construction Grant Amendment	1981	Reduction of federal share of municipal treatment plant funding to 55 percent
Clean Water Act Reauthorization	1987	Increased focus on non point source pollution
		Federal support for municipal treatment plants construction converted from grants to loans

2. **The economic impact of the Clean Water Law.**

   Environment regulations in most cases provides economic and health benefits that to a great extent outweighs the cost of compliance therefore leading to creation of employment. Since the various environmental laws have been enacted the United State gross domestic product has risen by over 200%. according to (Orth, 2013) Clean water alone is approximated to provide $11 billion annual benefits. The act has further enhanced the restoration of most of the water sourses such as the lakes and seas therefore resulting to increased recreational opportunities. For example in beaches where the dumping as intially dumped they have been restored thus more sites for tourists attraction both domestic and foreign (Mayer, 2012).

   The Act has resulted to increased supply of quality water to people these has increase productivity of because of increased health at the same time cost of health care has significantly reduced. With the efforts of implementing the Act new technologies has been introduced and more coming all with the efforts of implementing the Act. These growth in technology plays a biggest role in the growth and development of economy as modern infrsturcture are constructed.

3. **Improvement of Clean Water Act on the environment or situation.**

   The Clean Water Act has been one of the greatest revelations and most successful story of environmental law. Heavy investment was facilitated through enactment of this Act. Sewage treatment plants were constructed and those existing were upgraded. The Clean Water Act has lead to a significance improvement in water quality. Gallons of raw sewage that was initially dumped in the sources of water have since been cleaned and restored to their initial state (Martins, 2014).

   The US Environmental Protection Agency recently did a review of the benefits of Clean Water Act to the National environment and from the results there has been a significant progress made in the improvement of water quality and some other environmental resources. There are visible environmental benefits at different site across the nation located in large urban-industrial sectors on major waterways. They all depict an improvement in the water quality and related resources in the environment following the passage and implementation of the act (Hawkins, 2013).

Environmental report released in 1997 shows that the number of rivers, estuaries safe for fishing and swim and lakes doubled within a period of 26 years since the first enactment of the Clean Water Act in 1972.

4. Do you think that sound science has proven that global warming is a credible threat or not?

For a number of years now, it has been in the news headlines about how the ice caps are melting along with other climatic changes because of global warming. With that in mind one would ask how about all this warm weather we have been experiencing? Just for the purpose of being facetious. From the recent reports from NASA and other scientific observers it shows that there is still more to learn about the Universe and its environmental changes. I contend that the "sky is falling" attitude by many from the scientific world regarding the global warming because they have not put into perspective history on the various climatic changes that have been experienced not only in historic times but also for decades ago.

Recent study by the US National Snow and Ice Data Center shows that the sea ice surrounding the Antarctic continent had reached a maximum of almost eight million square miles, which is far above the 1981 to 2010 average. On the other hand ice cap knows as the Arctic, to the northern of USA and Canada, its approximately is 2.04 million square miles. The recent low monthly average of 637,000 square miles occurred in 2012. Many scientists have joined in the debate on global warming by insisting that the theory of manmade climate is no longer scientifically credible. For example in an open letter to the Intergovernmental Panel on Climate Change (IPCC), Coleman wrote, "The Ocean is not rising significantly the polar ice is increasing and polar bears are growing in number." Coleman added, "Heat waves have actually decreased, not increased. (Glen, 2015)" These views are based on the findings of an international non-governmental body of scientists aimed at offering an 'independent second opinion of the evidence reviewed by the IPCC.'

The issues of global warming in today's scenario have become a political and environment agenda item, but the science is no longer valid. Even the National Oceanic and Atmospheric Administration in a recent released temperature data report it say that the United States has been cooling for the past decade (Glen, 2015).

Many scientists all over the world, studying various aspects of the global warming have overestimated the climate change predictions because they don't take into account how plants absorb carbon dioxide.

Another study done on Global Warming has indicated that bio-fuels made from leftovers of harvested corn plants are worse than gasoline claiming that bio-fuels release 7 percent more greenhouse gasses as compared to conventional gasoline. Lastly like so many studies being made by a multitude of scientists and researchers around the world, it's crucial to invest in studies and research needs to be made in an effort to learn more about the Earth and Nature in general (Glen, 2015).

5. **Should the United States adopt additional policies or laws to curb greenhouse gas emissions?**

Environmental Protection Agencies should establish standards for greenhouse gas emission so as to control different sources such as mobile and stationary sources under the clean Air Act. Some of the actions complted and proposals to implement Clean Air Act requirements for carbon pollution and other green house gases include

i. Carbon pollution standards for the power plants this proposal will cut carbon pollution from existing power plants.
ii. EPA and National highway Traffic Safety Administration have take steps together in encouraging production of a new generation of clean vehicles. This proposal is aims at saving more than five million barrels of oil through 2025 while reducing more millions metric tons of carbon dioxide released into the atmosphere.
iii. Renewable fuel standard program – these regulates transportation fuel sold in the United State ensuring it contains a minimum volume of renewable fuels.
iv. Landfill Air Pollution Standards these would require various landfills to capture additional landfill gas which will reduce emissions of methane gases.

6. Conclusion

Despite the achievements made over the years since the introduction of Clean Water Act and various amendments made through the time there still exists significant problems. Water sources are still being polluted due to rise increased sources of pollution for example agricultural pollutions from use of fertilizers, manure and pesticides. Also there is difficulty in controlling of non point source of pollution for example water running off buildings, acid rain, and urban storm water, construction run off and run off from the mines. A lot of research needs to be done so as to come up with effective strategies that will help mitigate water pollution so that to restore and maintain sustainable water resources.

7. Works Cited

Glen, D. (2015, January 5). *Credibility of Global Warming Threat.* Retrieved February 7, 2015, from Your Houston News:

http://www.yourhoustonnews.com/cleveland/opinion/dodson-global-warming-threat-no-longer-credible/article_ad99afc7-175a-5fab-9738-bf38095ce80d.html

Hawkins, G. S. (2013). Clean Water Act. *A Journal of Ideas* , 76-89.

History of the clean water. (2013, March 21). Retrieved February 7, 2015, from Environmental Protection Agency: http://www2.epa.gov/laws-regulations/history-clean-water-act

Martins, T. (2014). What has the Clean Water Act Accomplished. *Journal of Environmental Policies* , 45-52.

Mayer, S. M. (2012, June 30). *The Economic Impact of Environmental Regulation* . Retrieved February 7, 2015, from Bechtel Colorado Education:

http://bechtel.colorado.edu/~silverst/cven5534/Economic%20Impact%20Environ.%20Regulation.pdf

Orth, J. (2013, November 19). *Clean water is good for the economy.* Retrieved February 7, 2015, from St Jonhs River Keeper: http://www.stjohnsriverkeeper.org/blog/clean-water-act-is-good-for-our-economy/

Regulatory initiatives. (2011, May 23). Retrieved January 7, 2015, from US Environmental Protection Agency: http://www.epa.gov/climatechange/EPAactivities/regulatory-initiatives.html

YOUR KNOWLEDGE HAS VALUE

- We will publish your bachelor's and master's thesis, essays and papers

- Your own eBook and book - sold worldwide in all relevant shops

- Earn money with each sale

Upload your text at www.GRIN.com
and publish for free